Salma Messous

Bacteriology of acute COPD exacerbations in Tunisia

Salma Messous

Bacteriology of acute COPD exacerbations in Tunisia

Imprint

Any brand names and product names mentioned in this book are subject to trademark, brand or patent protection and are trademarks or registered trademarks of their respective holders. The use of brand names, product names, common names, trade names, product descriptions etc. even without a particular marking in this work is in no way to be construed to mean that such names may be regarded as unrestricted in respect of trademark and brand protection legislation and could thus be used by anyone.

Cover image: www.ingimage.com

This book is a translation from the original published under ISBN 978-620-2-28360-1.

Publisher:
Sciencia Scripts
is a trademark of
Dodo Books Indian Ocean Ltd. and OmniScriptum S.R.L publishing group

120 High Road, East Finchley, London, N2 9ED, United Kingdom
Str. Armeneasca 28/1, office 1, Chisinau MD-2012, Republic of Moldova, Europe
Printed at: see last page
ISBN: 978-620-5-93945-1

Table of contents

Chapter 1

I.   Introduction

Chronic obstructive pulmonary disease (COPD) is one of the most common diseases in the world. It was the 3ème leading cause of death in 2010 and a frequent reason for admission to emergency departments and intensive care units [1,2].

An acute exacerbation of COPD (AECOD) is an exaggeration of the symptoms of chronic bronchitis. Currently, Anthonisen's three criteria [3] appear to be the most satisfactory for defining an APECO: increased sputum volume, change in sputum appearance to purulent, and increased dyspnoea. Infectious causes of exacerbation have probably been under-analysed [4]. Since the revision of the recommendations of the French Health Products Safety Agency (Afssaps), there are very few publications to report on COPD and its infectious causes. In general, in 30% of cases, the cause of EABPCO is bacterial, including *Chlamydophila pneumoniae* and *Mycoplasma pneumoniae*. In 25% of cases, it is viral in origin. It is both viral and bacterial in 25% of cases and is not found in 20% of cases [5,6]. Non-infectious causes are also responsible for exacerbations, such as active or passive smoking and exposure to other household or occupational irritants [5]. Cytobacteriological analysis of bronchial secretions is not routinely performed in clinical practice in COPD. Its indications and modalities vary from one recommendation to another. Furthermore, the distinction between infection and colonisation is still far from being resolved. It would be interesting and topical to discuss the presence of a microbiome in the branches. Since in healthy subjects the bronchial tree and lung parenchyma are considered sterile, potentially pathogenic microorganisms (PPM) are often recovered from bronchial secretions in stable COPD and especially in exacerbating COPD with a higher bacterial load [7]. These results confirm the colonisation of PBMs in the bronchial tree in COPD patients. On the other hand, *Chlamydophila pneumoniae* and *Mycoplasma pneumoniae* are common human pathogens causing asymptomatic or severe upper and lower respiratory tract infections with significant seroprevalence in the general population [6,8]. Very few publications exist on other atypical germs such as *Coxiella burnetii* and *Legionella pneumophila* [9]. Although COPD is a common cause of mortality worldwide, most of the microbiological research on this topic comes from developed countries and few such data exist for the Middle East and North Africa [10]. Estimates of bacterial ecology are also highly variable. In Tunisia, this prevalence remains unspecified. Hence the need for this work on the etiology of infectious causes in EABPCO by combining classical culture and serology of atypical germs. A precise definition of the bacteria in EABPCO is important to understand and guide the antimicrobial therapeutic approach. Especially as there are non-fermentative Gram-negative bacilli such as *Pseudomonas aeruginosa* and *Acinetobacter baumannii* that are reported to be multi-drug resistant (MDR) [11].

The aim of this study was to describe the bacterial flora, including the detection of atypical germs, in patients admitted to the emergency department for COPD.

II. Patients and Methods

2.1. . Design of the study

2.1.1. . Type of study

This is a multicentre descriptive and analytical study, conducted between March 2013 and May 2016 at four participating centres in Tunisia: Fattouma Bourguiba University Hospital in Monastir, Farhat Hached University Hospital in Sousse, Taher Sfar University Hospital in Mahdia, and Moknine Regional Hospital.

2.1.2. . Inclusion criteria

All patients of both sexes over 40 years of age, hospitalised with COPD and defined by increased dyspnoea, sputum volume and sputum purulence according to the New York Heart Association (NYHA) classification of dyspnoea were recruited [12] :

- I Dyspnoea for unusual heavy exertion: the patient has no discomfort in everyday life.

- II Dyspnoea for usual heavy exertion such as brisk or uphill walking, stair climbing (> 2 floors).

- III Dyspnoea for low intensity efforts of everyday life such as normal walking on flat ground, stair climbing (= 2 floors).

- IV Permanent rest dyspnoea.

To be included, patients must have polypnoea (>30 cycles/min) and/or acute hypercapnia on blood gas before initiation of mechanical ventilation (PaC02> 6kPa and arterial pH <7.30).

2.1.3. . Exclusion criteria

Patients with clinico-radiological evidence of acute pneumonia, fever (>38.5°C) or having received antibiotic treatment in the 10 days prior to sampling, and with other infectious diseases requiring antibiotic treatment or having been intubated at the outset were excluded.

The absence of radiological signs of pneumonia was confirmed in all patients by the investigating physician; a second chest X-ray was taken if necessary. Patients were also excluded if they had a history of allergy to quinolones; current pregnancy or breastfeeding; severe chronic disease: liver, kidney; known immunodeficiency (haematological malignancy, AIDS...); or a digestive disease that could affect drug absorption.

2.1.4. . Ethics

The study was approved by the ethics committee of the University Hospital of Sousse. Only patients who gave their free and informed consent to participate in the study were recruited.

2.2. . Conduct of the study

This study was part of a larger randomised double-blind clinical trial registered in the *ClinicalTrials.gov database as* NCT02067780 to compare two antibiotic treatment strategies for CAPCO: a short two-day course of Levofloxacin versus a standard seven-day course of the same drug, as an extension of a similar study previously conducted [13]. The patients included in the study were admitted to the emergency room intensive care unit. They were placed on oxygen and given non-invasive ventilation when indicated. If this therapy failed, patients were secondarily intubated and mechanically ventilated in the assist-control mode. Except for antibiotics, patients received instrumental and medical treatment according to current guidelines [14]. Associated medications included subcutaneous heparin and bronchodilators. All patients received enteral nutrition via oral or nasogastric tube or orogastric tube.

On admission, blood samples were routinely collected for all included patients for standard biological analysis, including blood gas, CBC (white blood cells and platelets), and CRP measurement. CRP concentrations were determined using an immunoturbidimetric assay (Roche Diagnostics, Indianapolis, IN, USA) consisting of agglutination of human CRP on latex particles coated with monoclonal anti-CRP antibody.

2.2.1. . Bacteriological examination of sputum (ECBC)

ECBC consists of immediate direct examination of the sputum for cell count and Gram stain for microorganisms, followed by culture, the results of which are not returned until 48-72 hours later.

On admission, sputum was collected in the morning in a sterile container after gargling with saline repeated three times to remove saliva and oropharyngeal secretions. The patient should be made aware of a coughing effort so that it brings back secretions from the deep lung. Purulent and mucopurulent sputum is suitable for bacteriological investigations, whereas salivary sputum should be rejected. The transfer to the microbiology laboratory of the University Hospital of Monastir was carried out in less than an hour.

The bacteriological examination of the sputum is carried out on the first day by a macroscopic examination to assess the appearance of the sample, which may be mucous, mucopurulent, purulent, haematic or salivary. A microscopic examination is carried out after Gram staining to determine the microbial flora and to carry out a cytology (count of bronchial cells, polynuclear cells etc.). This examination was an essential step in the diagnosis because, not only did it make it possible to judge the acceptability of the sample, but it was also used to guide the investigations thanks to the appearance of the flora. Only sputum that met the Bartlett cytological criteria adopted by Murray and Washington were cultured: epithelial cells less than or equal to 10 and leukocytes greater than or equal to 25 per microscopic field at 40x magnification. The results of this microscopic examination allow five classes of sputum to be distinguished **(Table I)**: **Table I.** Classification of sputum according to the degree of contamination.

Classes (Score Murray and Washingt on)	(x100) Cells/fields		Gram appearance	Indication for cultivation (Bartlett)	Results
	Epithelial	Leukocytes			
1	>25	<10	Regardless of the Gram	No	Strong oral contamination not allowing an objective interpretation of the examination
2	>25	10-25	Regardless of the Gram	No	Strong oral contamination not allowing an objective interpretation of the examination
3	>25	>25	Polymorphic bacterial flora	No	Strong oral contamination not allowing an objective interpretation of the examination
4	10-25	>25	Regardless of the Gram	**Yes**	Seed
5	<10	>25	Regardless of the Gram	**Yes**	Seed

A culture was considered significant when the dominant flora exceeded the quantitative threshold of 10^7 colony forming units (cfu) per millilitre. Bacteria were identified according to the standard recommendations of the French Society of Microbiology [15]. Sputum isolates were classified as either potential pathogens including *H. influenzae*, *B. catarrhalis*, *S. pneumoniae*, *Pseudomonas aeruginosa, Staphylococcus aureus*, and other gram-negative bacilli or as normal flora.

2.2.2. . Atypical germ serology

Four ml of the blood sample should be taken on admission (day 0) and after 15 days for the detection

of atypical germs: *Mycoplasma pneumoniae, Coxiella burnetii, Chlamydophila pneumoniae* and

Legionella pneumophila by indirect immunofluorescence and enzyme-linked immunosorbent assay.

a) - The indirect immunofluorescence method

The indirect immunofluorescence method is based on the reaction of antibodies in the sample with

the antigen bound to the surface of the slide. The specific antibodies present in the sample react with

the antigen and the immunoglobulins, which are not bound by the reaction with the antigen, are

removed in the rinsing process. In the next step, the antigen-antibody complex is revealed by the

FITC conjugate and can be visualised using the fluorescence microscope.

Serum IgG and IgM antibodies against *Mycoplasma pneumoniae* were measured by indirect

immunofluorescence technique (EUROIMMUN Medizinische Labordiagnostika AG, Lübeck,

Germany) according to the manufacturer's instructions. This test is designed exclusively for the in

vitro determination of human antibodies in serum or plasma. The determination can be performed

qualitatively or quantitatively. Infected and uninfected *M.pneumoniae* cells are incubated with the

diluted sample. If the reaction is positive, specific antibodies of the IgG and IgM class bind to the

bacterial antigens. In a second step, the bound antibodies are stained with fluorescein-labelled anti-

human antibodies and made visible with a fluorescence microscope.

Detection of IgG and IgM antibodies against *Coxiella burnetii* was performed according to the

instructions of the indirect immunofluorescence kit (PCOBU I+II, Vircell, Spain).

b)- The enzyme-linked immunosorbent assay (ELISA)

The Enzyme Linked ImmunoSorbent Assay (ELISA) is an enzyme-linked immunosorbent assay that

visualises an antigen-antibody reaction by means of a coloured reaction produced by the action of an

enzyme on a substrate previously attached to the antibody. Detection of IgG and IgM antibodies to

Chlamydophila pneumoniae was performed using the ELISA technique (NovaTec,

Immunodiagnostica GmbH, Dietzenbach, Germany) according to the manufacturer's instructions. The wells of the microtiter strips are coated with *C. pneumoniae* antigens to bind the corresponding antibodies of the specimen. After washing the wells to remove the detached sample, horseradish peroxidase (HRP) is added to label the anti-IgM or IgG conjugate. This conjugate binds to antibodies specific to the captured *C. pneumoniae*. The immune complex formed by the bound conjugate is visualised by the addition of tetramethylbenzidine (TMB) which gives a blue reaction product. The intensity of this product is proportional to the amount of *C. pneumoniae* specific IgM or IgG antibody in the sample. Sulphuric acid is added to stop the reaction. This produces a yellow end colour. The absorbance at 450 nm is read using an ELISA microplate reader.

Detection of IgG and IgM antibodies to *Legionella pneumophila* was performed with the ELISA technique (EUROIMMUN Medizinische Labordiagnostika AG, Lubeck, Germany) according to the manufacturer's instructions. This kit provides a semi-quantitative in vitro analysis for human IgM or IgG antibodies to *L. pneumophila* in serum or plasma. It contains microtitre strips each with 8 wells of breakable reagents coated with *L. pneumophila* antigens. In the first step of the reaction, diluted samples are incubated in the wells. In the case of a positive sample, specific IgM or IgG antibodies bind to the antigens. To detect the bound antibodies, a second incubation is performed using an enzyme-labelled anti-human IgM or anti-IgG (enzyme conjugate) that catalyses a colour reaction. These serological tests were performed in the microbiology laboratory of the University Hospital of Monastir. The rules of interpretation of our results are as follows:

- A patient was considered not to be infected with such an atypical bacterium if IgM and IgG antibodies were absent in both sera.

- A patient with negative IgM and positive IgG on both sera was considered to have old immunity to such an atypical bacterium.

- a patient was considered to have an acute infection with such an atypical bacterium if IgM and IgG antibodies were present in the admission sample on day 0 or if seroconversion occurred between days 0 and 15.

- An acute reactivation pattern is a seroconversion of IgG without IgM.

- Serum samples showing IgM at day 0 without IgG at day 15 were considered as false positive results.

2.3. . Patient follow-up

Post-discharge follow-up at 6 months and 1 year was done for all included patients by simple telephone contact to record data on the occurrence of a new exacerbation.

2.4. . Statistical analysis

The data were entered into SPSS software version 18.0 after collection from the patient charts. The statistical analysis was carried out on the same software based on the drafting of a syntax corresponding to the plan of the results of our study. The description of the data was based on the presentation of the numbers and relative frequencies for the categorical variables, the mean and standard deviation for the continuous variables.

The assumption of normality was considered for all quantitative variables in the work. The comparison of percentages was tested by the Chi-square test, or its correction for continuity (Yates' or Fisher's Chi-square, if the condition of theoretical values > 5 *was* not verified). In addition, the comparison of means between groups was done by a one-factor analysis of variance. No specific adjustments were made, and statistical inference was based on two-tailed tests at the 5% significance level. A description of the data at inclusion was made for socio-demographic parameters, clinical data, history of disease and complementary investigations.

Chapter 3

III. Results

3.1. 1. Socio-demographic characteristics

Two hundred and forty patients were included in the study. The mean age was 68.3 ±10.5 with a predominance of males (90.8%). Almost half of the patients had severe COPD at NYHA stage 3 (48.8%) and stage 4 (44.2%). According to the clinical history of the exacerbation, 65 patients (39.4%) had one Anthonisen criterion, 57 patients (34.5%) had two Anthonisen criteria and 43 patients (26.1%) had all three Anthonisen criteria. More than half of the population were active smokers (66.6%) with an average annual tobacco consumption of 74.1±127 pack-years.

3.2. 2. Potentially pathogenic bacteria

Of the 240 patients included, 175 sputum cultures (73%) were considered significant. Class 1, 2 and 3 sputum are highly contaminated with saliva and should not be cultured. A new sample should be requested. Class 4 and 5 sputum have a leukocyte count that indicates an inflammatory reaction but are contaminated by saliva **(Table II).**

Table II.   Interpretation of the microscopic examination of the sputum of the included patients according to the recommendations of the medical microbiology reference of the French society of Microbiology [15].

(x100) Cells/fields		Score (Murray and Washington)	Indication for cultivation (Bartlett)	Number of samples
Epithelial	Leukocytes			
>25	<10	1	No	7
>25	10-25	2	No	0
>25	>25	3	No	1
10-25	<10	Not specified	No	0
10-25	10-25	Not specified	No	59
10-25	>25	4	**Yes**	7
<10	<10	Not specified	No	0
<10	10-25	Not specified	**Yes**	**0**
<10	>25	5	**Yes**	**168**
			Total	240

Twenty-nine cultures were positive (16.5%) and 31 germs were isolated. The 4 most frequently isolated species were *P. aeruginosa* (25.8%), *K. pneumoniae* (16.2%), *H. influenzae* (13.0%) and *S. pneumoniae* (9.7%) **(Table III).**

Table III.   Bacteriological profile of sputum cultures and serological results of included

patients.

	Isolated germs	Number of isolates N	Frequency %.
Potentially pathogenic microorganisms (PPM)	**MPP (n=175)** *Acinetobacter spp* ($>10^7$ CFU/ml)	1	3.2
	Branhamella catarrhalis ($>10^7$ CFU/ml)	1	3.2
	Haemophilus parainfluenzae ($>10^7$ CFU/ml)	1	3.2
	Proteus mirabilis ($>10^7$ CFU/ml)	1	3.2
	Providencia ($>10^7$ CFU/ml)	1	3.2
	Staphylococcus aureus ($>10^5$ CFU/ml)	1	3.2
	Acinetobacter baumanni ($>10^7$ CFU/ml)	2	6.4
	Escherichia coli ($>10^7$ CFU/ml)	3	9.7
	Streptococcus pneumoniae ($>10^7$ CFU/ml)	3	9.7
	Haemophilus influenzae ($>10^7$ CFU/ml)	4	13
	Klebsiella pneumoniae ($>10^7$ CFU/ml)	5	16.2
	Pseudomonas aeruginosa ($>10^7$ CFU/ml)	8	25.8
Atypical germs			
	Chlamydophila pneumoniae (n=166)	14	8.4
	Mycoplasma pneumoniae (n=166)	15	9.0
	Coxiella burnetii (n=91)	6	6.6
	Legionella pneumophila (n=91)	0	0
	Non-MPP	146	83.4
Coinfections	**Coinfections (n=10)**		
	Acinetobacter spp / Providencia	1	10
	Klebsiella pneumoniae / Pseudomonas aeruginosa	1	10
	Klebsiella pneumoniae /Haemophilus influenzae	1	10
	Klebsiella pneumoniae /Proteus mirabilis	1	10
	Escherichia coli / Chlamydia pneumoniae	1	10
	Haemophilus influenzae / Chlamydia pneumoniae	1	10
	Pseudomonas aeruginosa / Coxiella burnetii	1	10
	Pseudomonas aeruginosa /Mycoplasma pneumoniae	2	20
	Streptococcus pneumoniae / Chlamydia pneumonia	1	10

.2.1.   Distribution of enterobacteria isolated from sputum

Ten Enterobacteriaceae were found in our study sample, with *Klebsiella pneumoniae* being the
most common (50%) **(Figure 1).**

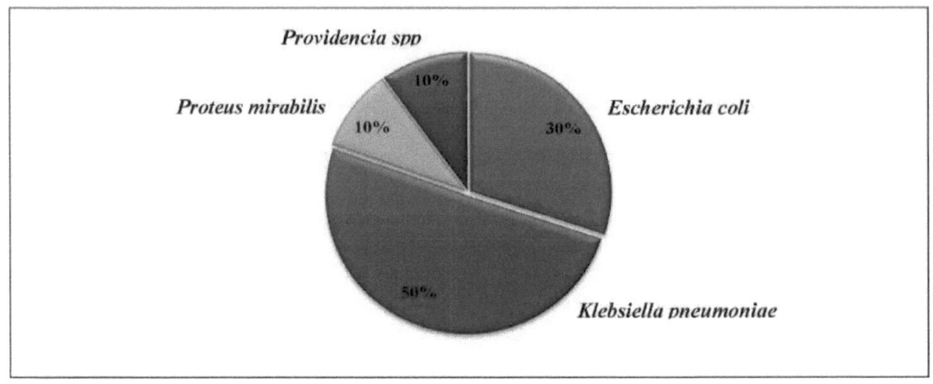

Figure 1: Distribution of Enterobacteriaceae isolated according to bacterial species

All species of *K. pneumoniae* were extended-spectrum beta-lactamase (ESBL) producers except for one low-level penicillinase producer.

111.2.2. Distribution of non-enterobacteria isolated from sputum
Eighteen non-enterobacteria were found in the samples studied, of which pyocyanine bacillus was found in the majority (50%) **(Figure 2)**.

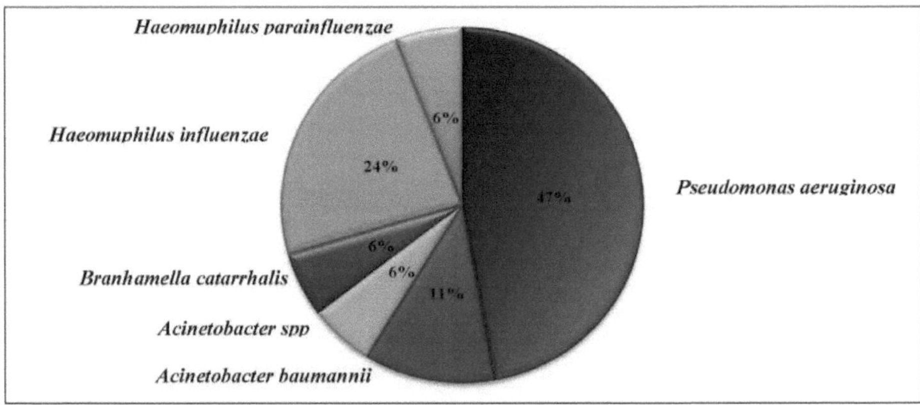

Figure 2: Distribution of non-enterobacteria by bacterial species

Four samples with the following associations:
- *Acinetobacter spp* and *Providencia* (n=1),

- *K. pneumoniae* and *P. aeruginosa* (n= 1),

- *K. pneumoniae* and *P. mirabilis* (n=1),

- *K. pneumoniae* and *H. influenzae* (n=1).

III .3. Antibiotic susceptibility testing

Almost half of the strains tested were resistant to the first-line antibiotics recommended by the learned societies. Resistance to aminopenicillin antibiotics was predominant [70.5% for Amoxicilin and 43.7% for Amoxicillin-Clavulanic acid]. Resistance to fluoroquinolones was less important with 25% for Levofloxacin. For *Haemophilus*, the bacteriological situation is comparable to that of the pneumococcus with a slightly lower degree of resistance to the classical antibiotics in question (**Table IV**).

Table IV. General resistance profile of all isolated strains to the antibiotics tested.

Antibiotic tested / Strain tested	AMX	AMC	GM	CS	OFX	CIP	LVX	RA	SXT	Total
Acinetobacter spp	R	R	^B	^B	^B	R	-	-	-	
Acinetobacter baumanni	R	R	R	S	R	R	-	-	R	
Acinetobacter baumanni	R	R	S	S	^B	R	-	-	-	
Branhamellacatarrhalis	R	S	S	-	^B	S	^B	S	-	
Escherichia coli	R	R	S	S	S	^B	-	S	S	
Escherichia coli	-	-	-	^B	^B	-	-	-	-	
Escherichia coli	R	S	S	S	R	R	-	-	R	
Haemophilus influenzae	S	S	S	-	-	-	^B	S	-	
Haemophilus influenzae	S	S	-	^B	^B	-	-	-	-	
Haemophilus influenzae	S	S	R	-	^B	R	^B	S	S	
Haemophilus parainfluenzae	-	S	^B	^B	^B	S	-	-	-	
Pseudomonas aeruginosa	-	-	S	S	^B	S	-	-	-	
Pseudomonas aeruginosa	-	-	S	S	^B	S	-	-	-	
Pseudomonas aeruginosa	-	-	S	S	^B	R	-	-	-	
Pseudomonas aeruginosa	-	-	R	-	^B	S	-	-	-	
Pseudomonas aeruginosa	-	-	R	S	^B	R	-	-	-	
Pseudomonas aeruginosa	-	-	R	-	^B	R	-	-	-	
Pseudomonas aeruginosa	-	-	^B	S	^B	S	-	-	-	
Pseudomonas aeruginosa	-	-	S	S	^B	R	-	-	-	
Staphylococcus aureus	-	-	-	-	S	S	S	S	-	

Strain										
Streptococcus pneumoniae	-	-	R	-	-	S	S	S	R	
Streptococcus pneumoniae	S	^B	R	-	-	IT	R	S	R	
Streptococcus pneumoniae	S	^B	R	-	-	IT	S	-	-	
Klebsiella pneumoniae	R	IT	S	S	R	R	-	R	R	
Klebsiella pneumoniae	R	IT	R	S	S	S	-	R	S	
Acinetobacter spp /Providencia	R	R	S	S		R	-			
Klebsiella pneumoniae / Pseudomonas aeruginosa	R	R	R	S	S	S	-	R	S	
Klebsiella pneumoniae / Haemophilus influenzae	R	S	S	S	S	S	-	R		
Klebsiella pneumoniae / Proteus mirabilis	R	R	R	S	S	S	-	IT	S	
Number of strains tested	17	16	23	16	9	25	4	12	10	**132**
Number of resistant strains	12	7	11	0	3	11	1	4	5	**54**
of resistant strains	70.5	43.7	47.8	0	33.3	44	25	33.3	50	**40.9**

R: resistant; **S**: susceptible; **IT**: Intermediate; **AMX**: Amoxicillin; **AMC**: Amoxicillin+AC. Clavulanic; **GM**: Gentamicin; **CS**: Colistin; **OFX**: Ofloxacin; **CIP**: Ciprofloxacin; **LVX**: Levofloxacin; **RA**: Rifamicin; **SXT**: Trimethoprim+Sulfonamides; **P**: Penicillin G.

Table V shows the evolution of the sensitivity of isolated Enterobacteriaceae to antibiotics.

Table V.   Antibiotic resistance profile of Enterobacteriaceae

Strain / Antibiotics	*E. coli* (N=3)			*K. pneumoniae* (N=5)			*P. mirabilis* (N=1)			*Providencia spp* (N=1)		
	R	%	NT	R	%	NT	R	%	NT	R	%	NT
AMOXICILLIN	2/2	100	1	5/5	100	0	1/1	100	0	1/1	100	0
AMOXICILLIN + AC.CLAVULANIC	2/2	100	1	2/3	66,66	2	1/1	100	0	1/1	100	0
TICARCILLIN	1/2	50	1	5/5	100	0	1/1	100	0	1/1	100	0
TICARCILLIN + AC.CLAVULANIC	1/2	50	1	3/4	75	1	0/0	0	1	1/1	100	0
PIPERACILLE	0/1	0	2	5/5	100	0	1/1	100	0	-	-	-
PIPERACILLIN + TAZOBACTAM	0/2	0	1	0/5	0	0	0/1	0	0	0/0	0	1
IMIPENEME	0/2	0	1	0/5	0	0	0/1	0	0	0/1	0	0
CEFOTAXIME	0/2	0	1	4/5	80	0	1/1	100	0	1/1	100	0
CEFALOTINE	1/2	50	1	4/5	80	0	1/1	100	0	1/1	100	0
CEFAMANDOLE	0/2	0	1	4/5	80	0	1/1	100	0	-	-	-
CEFOXITINE	0/2	0	1	2/5	40	0	0/1	0	0	0/0	0	1
CEFIXIME	-	-	-	2/2	100	3	1/1	100	0	-	-	-
CEFTAZIDINE	0/2	0	1	4/5	80	0	1/1	100	0	1/1	100	0
CEFPIROME	0/2	0	1	1/2	50	3	1/1	100	0	0/1	0	0
MECILLINAM	0/1	0	2	4/4	100	0	1/1	100	0	-	-	-
GENTAMYCINE	0/2	0	1	3/5	60	0	0/1	0	0	1/1	100	0
COLISTINE	0/2	0	1	0/5	0	0	1/1	100	0	1/1	100	0
TETRACYCLINE	0/2	0	1	2/5	20	0	1/1	100	0	1/1	100	0
AZTREONAM	0/2	0	1	1/1	100	4	-	-	-	0/1	0	0
ERTAPENEME	0/2	0	1	1/5	0	0	0/1	0	0	0/1	0	0
TRIMETHOPRIM + SULFAMIDES	1/2	50	1	1/5	20	0	0/1	0	0	1/1	100	0

OFLOXACINE	0/2	0	1	1/5	20	0	0/1	0	0	-	-	-
NORFLOXACIN	0/1	0	2	1/5	20	0	0/1	0	0	-	-	-
FOSFOMYCINE	-	-	-	0/1	0	4	-	-	-	1/1	100	0
AMIKACINE	0/2	0	1	3/4	75	1	0/1	0	0	-	-	-
TOBRAMYCINE	-	-	-	4/5	80	0	0/1	0	0	1/1	100	0
NALIDIXIC ACID	0/1	0	2	0/3	0	0	0/1	0	0	-	-	-
NETILMICIN	-	-	-	-	-	-	-	-	-	1/1	100	0
RIFAMPICIN	0/1	0	2	4/4	100	1	0/0	0	1	-	-	-
FURANES	0/1	0	2	0/2	0	3	1/1	100	0	-	-	-
TIGECYCLINE	-	-	-	1/2	50	3	-	-	-	-	-	-

All Enterobacteriaceae expressed resistance to Amoxicillin (100%) **(Figure 3)**.

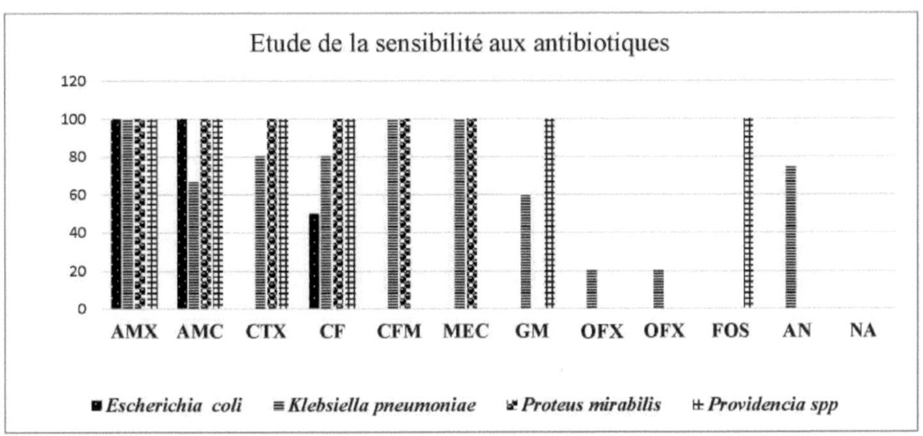

Figure 3. Resistance profile of isolated Enterobacteriaceae

AMX: Amoxicillin; **AMC**: Amoxicillin +Ac. Clavulanique; **CTX**: Cefotaxime; **CF**: Cefalotine; **CFM**: Cefixime; **MEC**: Mecillinam; **GM**: Gentamicin; **OFX**: Ofloxacin; **NOR**: Norfloxacin; **FOS**: Fosfomycin; **AN**: Amikacin; **NA**: Naldixic acid Table VI shows the evolution of the susceptibility of non-enterobacteria isolated to antibiotics.

Table VI. Resistance profile of non-enterobacteria to antibiotics

	P. aeruginosa (N=9)			H. influenzae (N=3)			A. baumanni (N= 2)			A. spp (N= 1)			B. catarrhalis (N=1)		H parainfluenzae (N=1)	
	R	%	NT	R	%	NT	R	%	NT	R	%	NT	R	%	R	%
AMOXICILLIN	-	-	-	0/3	0	0	2/2	100		1/1	100	0	1/1	100	-	-
AMOXICILLIN + AC.CLAVULANIC	-	-	-	0/3	0	0	2/2	100		1/1	100	0	0/1	0	0/1	0
TICARCILINE	5/9	55,5	0	-	-	-	1/1	100	1	0/1	0	0	-	-	-	-
TICARCILINE + AC.CLAVULANIC	3/9	33,3	0	-	-	-	1/1	100	1	0/1	0	0	-	-	-	-
PIPERACILLIN	4/9	44,4	0	-	-	-	1/1	100	1	-	-	-	-	-	-	-
PIPERACILLIN + TAZOBACTAM	2/9	22,2	0	-	-	-	1/1	100	1	-	-	-	-	-	-	-
CHLORAMPHENICOL	-	-	-	1/3	33	0	-	-	-	-	-	-	0/1	0	0/1	0
CEFOTAXIME	-	-	-	0/1	0	2	1 /1	100	1				1/1	100	0/1	0
CEFALOTINE	-	-	-	1 /2	50	1	1/1	100	1	1/1	100		-	-	-	-

CEFSULODINE	1 /2	50	7	-	-	-	-	-	-	-	-		-	-	-	-
CEFTAZIDINE	3/7	42,8	2	-	-	-	2/2	100	0	0/1	0		-	-	-	-
CEFOXITINE	-	-	-	0/1	0	2	2/2	100	0	1/1	100		-	-	-	-
CEFAMANDOLE	-	-	-	-	-	-	1/1	100	1	-	-		-	-	-	-
IMIPENEME	2/9	22,2	0	-	-	-	1/1	100	1	0/1	0		-	-	-	-
RIFAMPICIN	-	-	-	0/1	0	2	-	-	-	-	-		0/1	0	-	-
ERYTHROMYCIN	-	-	-	0/1	0	2	-	-	-	-	-		0/1	0	0/1	0
AZITHROMYCIN	-	-	-	0/1	0	2	-	-	-	-	-		-	-	0/1	0
KANAMYCINE	-	-	-	0/2	0	1	-	-	-	-	-		0/1	0	0/1	0
CEFEPIME	4/9	44,4	0	-	-	-	1 /1	100	1	0/1	0		-	-	-	-
GENTAMICIN	3/8	37,5	1	0/3	0	0	1/1	100	1	0/1	0		0/1	0	0/1	0
AZTREONAM	0/3	0	6	0/1	0	2	1/1	100	1	1/1	100		-	-	-	-
TETRACYCLINE	-	-	-	0/3	0	0	2/2	100	0	-	-		0/1	0	-	-
NALIDIXIC ACID	-	-	-	0/3	0	0	-	-	-	-	-		-	-	0/1	0
CIPROFLOXACIN	3/8	37,5	1	-	-	-	2/2	100	0	0/1	0		0/1	0	0/1	0
AMIKACINE	1/7	14,2	2	-	-	-	1 /2	50	0	-	-		-	-	-	-
COLISTINE	0/6	0	3	-	-	-	0/1	0	1	0/1	0		-	-	-	-
TRIMETHOPRIM + SULFAMIDES	-	-	-	0/1	0	2	1/1	100	1	0/1	0		-	-	-	-
FOSFOMYCINE	3/8	37,5	1	-	-	-	1/1	100	1	1/1	100		-	-	-	-
TOBRAMYCINE	2/8	25	1	-	-	-	1/1	100	1	-	-		-	-	-	-
ERTAPENEME	-	-	-	-	-	-	2/2	100	0	1/1	100		-	-	-	-
OFLOXACINE	-	-	-	-	-	-	1/1	100	1	-	-		-	-	-	-
NORFLOXACIN	-	-	-	-	-	-	1/1	100	1	-	-		-	-	-	-

A. baumannii strains expressed resistance to most of the antibiotics tested, whereas *P. aeruginosa* strains showed less resistance patterns **(Figure 4)**.

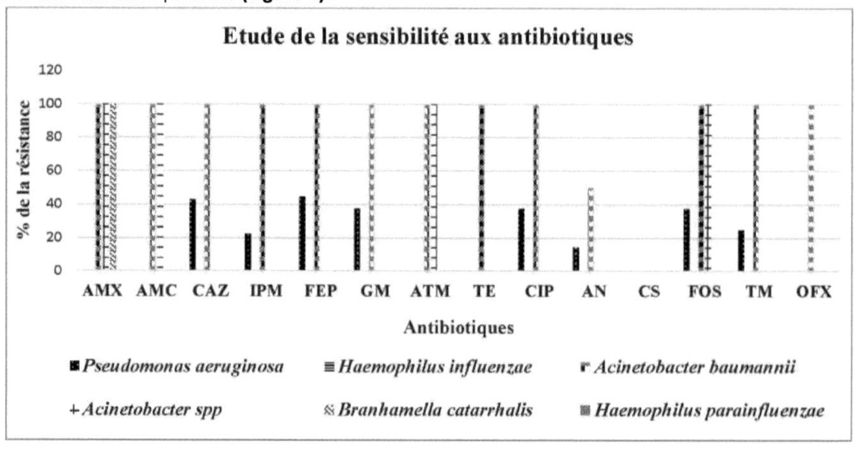

Figure 4.   Resistance profile of non-enterobacteria

AMX: Amoxicillin; **AMC**: Amoxicillin +Ac. Clavulanic; **CAZ**: Ceftazidine; **IPM** Imipenem; **FEP**: Cefepin; **GM**: Gentamicin; **ATM**: Aztreonam; **TE**: Tetracycline; **CIP** Ciprofloxacin; **AN**: Amikacin; **CS**: Colistin; **FOS**: Fosfomycin; **TM**: Tobramycin; **OFX** Ofloxacin.

Table VII shows the evolution of the sensitivity of isolated *Staphylococcus aureus* to antibiotics.

Table VII. Resistance profile of isolated *Staphylococcus aureus* to antibiotics

Strain Antibiotics	*Staphylococcus aureus* N= 2		
	R	**%**	**NT**
PENICILLIN G	1/2	**50**	0
CEFOXITINE	0/2	0	0
OXACILLIN	0/1	0	1
KANAMCYNE	0/2	0	0
TOBRAMYCINE	0/1	0	1
GENTAMICIN	0/1	0	1
AMIKACINE	0/2	0	0
STREPTOMYCINE	0/2	0	0
TETRACYCLINE	1/1	**100**	1
CHLORAMPHENICOL	0/2	0	0
ERYTHROMYCIN	0/2	0	0
LINCOMYCINE	0/2	0	0
SPIRAMYCINE	0/2	0	0
PRITINAMYCIN	0/2	0	0
OFLOXACINE	0/2	0	0
NORFLOXACIN	0/1	0	1
CIPROFLOXACIN	0/1	0	1
LEVOFLOXACINE	0/1	0	1
FUSIDIC ACID	0/2	0	0
TRIMETHOPRIM + SULFAMIDES	0/1	0	1
FOSFOMYCINE	0/1	0	1
VANCOMYCINE	0/2	0	0
TEICOPLANINE	0/2	0	0

Staphylococcus aureus strains show resistance to tetracycline (100%) and penicillin G (50%)

(Figure 5).

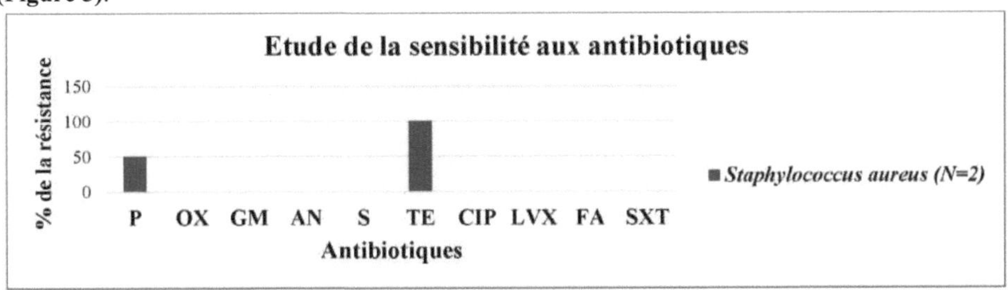

Figure 5.   Resistance profile of isolated *Staphylococcus aureus*

P: Penicillin G; **OX**: Oxacillin; **GM**: Gentamicin; **AN**: Amikacin; **S**: Streptomycin; **TE**: Tetracycline; **CIP**: Ciprofloxacin; **LVX**: Levofloxacin; **FA**: Fusidic acid; **SXT**: Trimethoprine + sulfamethoxazole.

Table VIII shows the evolution of the sensitivity of isolated *Streptococcus pneumoniae* to antibiotics.

Table VIII. Resistance profile of isolated *Streptococcus pneumoniae* to antibiotics

Strain Antibiotic^'"-\^^	*S. pneumoniae* (N=3)		
	R	%	NT
PENICILLIN G	0/2	0	1
AMOXICILLIN	0/3	0	0
CEFOTAXIME	0/3	0	0
GENTAMICIN HC	2/2	100	1
STREPTOMYCIN HC	3/3	100	0
KANAMYCINE HC	3/3	100	0
TETRACYCLINE	1/1	100	2
GENTAMYCINE	1/1	100	2
TOBRAMYCINE	1/1	100	2
NETILMYCINE	1/1	100	2
AMIKACINE	1/1	100	2
CHLORAMPHENICOL	0/3	0	0
ERYTHROMYCIN	3/3	100	0
LINCOMYCINE	3/3	100	0
PRISTINAMYCIN	0/3	0	0
NORFLOXACIN	2/3	66,6	0
CIPROFLOXACIN	0/1	0	2
LEVOFLOXACINE	1/3	33,3	0
VANCOMYCINE	0/3	0	0
TEICOPLANINE	0/2	0	1
TRIMETHOPRIM + SULFAMIDES	2/2	100	1
FOSFOMYCINE	0/1	0	2
RIFAMPICIN	0/2	0	1

The *Streptococcus pneumoniae* strains isolated showed resistance to the majority of antibiotics tested **(Figure 6)**.

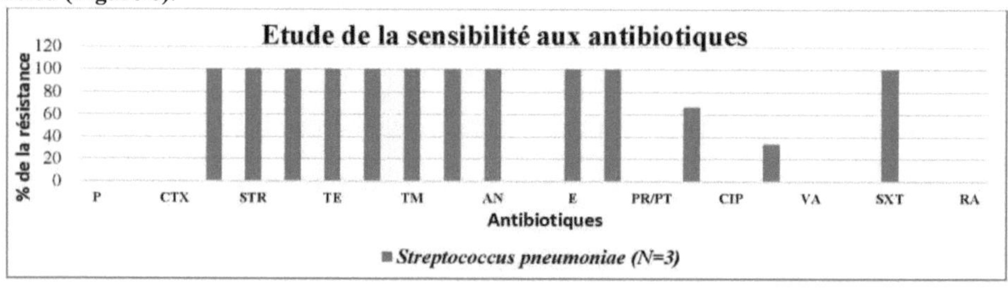

Figure 6. Resistance profile of isolated *Streptococcus pneumoniae*
P: Penicillin G; **AMX**: Amoxicillin; **CTX:** Cefotaxime; **GEN**: Gentamicin HC; **STR**: Streptomycin HC; **KAN**: Kanamycin HC; **TE**: Tetracycline; **GM**: Gentamycin; **TM**: Tobramycin; **NET**: Netilmycin; **AN**: Amikacin; **C**: Chloramphenicol; **E** : Erythromycin; **L**: Lincomycin; **PR/PT**: Pristinamycin; **NOR**: Norfloxacin; **CIP**: Ciprofloxacin; **LVX**: Levofloxacin; **VA**: Vancomycin; **TEC**: Teicoplanin; **SXT**: Trimethoprim + sulphamethoxazole; **FOS**: Fosfomycin; **RA**: Rifampin.

111.4. **Prediction of bacterial infection**

Culture positivity was not related to patient age, number of exacerbations per year, or smoking status.

No other clinical parameters were significant in predicting bacterial infection. However, the

Anthonisen criteria can be considered as partially predictive of a positive culture [44.4% of culture-positive patients with 3 Anthonisen criteria versus 22.5% of patients with 3 Anthonisen criteria but a negative culture] (p=0.04) **(Table IX)**.

111.5. Follow-up after hospital discharge at 6 months and 1 year

Follow-up of patients at 6 months and 1 year showed that patients with a positive culture had significantly more recourse to emergency readmission than those with a negative culture [6-month readmission: 41.3% versus 20.8%, p=0.02; 1-year readmission: 48.2% versus 27.4%, p=0.05] **(Table IX)**.

Table IX. Characteristics of patients admitted to the emergency department for an acute exacerbation of COPD.

Features	All patients	Culture		
	N= 240	negative N= 211	positive N=29	P
Demographic data				
- Age (years, mean ± SD)	68.3±10.5	68.2 ±10.7	69.5 ±8.8	NS
- Gender				NS
Male	218 (90.8)	190 (90)	28 (96.6)	
Woman	22 (9.2)	21(10)	1(3.4)	
- Active smokers	160 (66.6)	139 (65.8)	21 (72.4)	NS
- Smoking (pack year)	74.1±127	69.5±129	100.6±170	NS
-Annual number of CAPCOs (mean ± SD)	2.7±1.9	2.7±1.9	2.9±1.7	NS
Clinical data at admission				
- NYHA stage				NS
Stage 2		15 (7.1) 105	2 (6.9)	
Stage 3	17 (7.1) 117	(49.8)	12 (41.4)	
Stage 4	(48.8) 106 (44.2)	91 (43.1)	15 (51.7)	
-Clinical history of exacerbation: Anthonisen criteria Presence of one criterion Presence of two criteria Presence of three criteria	65 (39.3) 57 (34.5) 43 (26.1)	59 (42.8) 48 (34.8) 31 (22.5)	6 (22.2) 9 (33.3) 12 (44.4)	0.04
- Peak Expiratory Flow (L/min)	54.8 ± 73.8	53.9 ± 73.7	61.4 ±75.3	NS
- Body mass index (kg/m2)	25.8 ± 5.2	25.7 ±5.4	27.1 ±3.4	NS
- Body temperature (°C)	37.2 ± 0.6	37.1 ±0.6	37.1±0.6	NS
- Respiratory rate (c/min)	26.8 ± 7.9	26.3 ±5.9	30.2 ±15.7	0.01
- Heart rate (b/min)	108.6 ± 22.8	107.3 ±22.9	116.7 ±21.6	0.04
-Blood pressure				
Systolic (mmHg)	100 ± 76	97.8 ±78.4	120.5 ± 49.2	NS
Diastolic (mmHg)	65.5 ±99	61.7 ±86.2	93.4 ±163.2	NS
-**Intake ventilation mode**				
Ambient air	111 (46.2)	97 (45.9)	14 (48.2)	**NS**
Under O2	97 (40.4)	85 (40.2)	12 (41.3)	**NS**
Under NIV	32 (13.3)	29 (13.7)	3 (10.3)	**NS**
-**Background treatment**				
No	54 (22.5)	50 (23.7)	4 (13.7)	NS
Diuretic	8 (3.3)	6 (2.8)	2 (6.8)	NS

Converting enzyme inhibitors (CEIs)	42 (17.5)	36 (17)	6 (20.6)	NS
Beta-mimetic	58 (24.1)	51 (24.1)	7 (24.1)	NS
Inhaled corticosteroids alone	45 (18.8)	40 (18.9)	5 (17.2)	NS
Other	33 (13.8)	28 (13.2)	5 (17.2)	NS
Biological data on admission				
- PaO_2 (mmHg)	76.9 ± 32.2	77.5 ±32.5	72.7 ±30.8	NS
- PaCO2 (mmHg)	46.4 ± 18.2	47.3 ±18.8	40.4 ±12.6	NS
- pH	7.37 ± 0.07	7.37 ±0.07	7.36 ±0.07	NS
- Oxygen saturation (SaO)2	90.1 ± 10.3	90.7 ±9.9	86.2 ±11.7	0.03
- White Globule (c/mm^3 x 103)	11.7 ± 20.2	11.5 ±21.4	13.2 ±6.8	NS
- Hemoglobin (g/dL)	13.6 ± 3.3	13.6 ±3.4	13.4 ±2.1	NS
- Platelets (x10^3 /µL)	245± 94.8	247.5 ±91.9	226.8 ±113.7	NS
- C-Reactive Protein (mg/dL)	72.1 ±72	71.8 ±72.9	73.7 ±68.4	NS
Evolution before discharge from hospital				
-Satisfactory	148 (61.6)	134 (63.5)	14 (48.2)	NS

-Unsatisfactory
  Secondary use of NIV Use of intubation
  Death in hospital
	92 (38.3)
	31 (31.9) /n= 97 O2
	
	8 (8.6)
	6 (6.5)
		77 (36.4)
		26 (30.5) /n= 85 O2
		
		7 (9.0)
		5 (6.4)
			15 (51.7) 5
			(41.6) /n= 12 O2
			
			1 (6.6)
			1 (6.6)
				NS

What happens after discharge from hospital
  Readmission at 6 months
  Readmission at 1 year
	56 (23.3)
	72 (30)
		44 (20.8)
		58 (27.4)
			12 (41.3)
			14 (48.2)
				0.02
				0.05

p< 0.05 culture negative *versus* culture positive; **NS: Not Significant**

III .6. Atypical germs

Of the 240 EABPCO patients, only 91 were serologically tested for the atypical germs *M. pneumoniae, C. pneumoniae, C. burnetii* and L. *pneumophila* simultaneously. A further 75 patients were tested only for *M. pneumoniae* and *C. pneumoniae due to the* availability of kits. Patients with a single serum (n=72) or haemolytic specimens (n=2) were secondarily excluded from this study.

- IgM antibodies to *M. pneumoniae were* detected in only one patient. Fourteen other patients seroconverted for *M. pneumoniae* IgG antibodies without IgM antibodies, which could be consistent with a profile of acute reactivation. In summary, a possible or probable serological profile of acute *M. pneumoniae* infection was noted in 15 patients (9%).

- IgM antibodies to *C. pneumoniae* were detected in 5 patients, and 9 patients were also found to be positive with an acute reactivation pattern. In summary, a possible or probable serological profile of acute *C. pneumoniae* infection in early and late serum samples was noted in 14 patients (8.4%), while 52 had a profile of old immunity and 15 had no antibodies against this bacterium.

- IgM antibodies to *C. burnetii were* detected in 6 patients (6.6%) of whom only one patient had an acute reactivation pattern. While 7 other patients had no antibodies to this bacterium and 8

showed a long-standing immunity to this bacterium.

- No IgM or IgG detection was found for *L. pneumophila* **(Table III).**

Six patients had potentially pathogenic microorganism/atypical germ co-infections **(Table III)**:

- *Escherichia coli / Chlamydia pneumoniae* (n=1),

- *Haemophilus influenzae / Chlamydia pneumoniae* (n= 1),
- *Pseudomonas aeruginosa / Coxiella burnetii* (n=1),
- *Pseudomonas aeruginosa / Mycoplasma pneumoniae* (n=2).
- *Streptococcus pneumoniae / Chlamydia pneumoniae* (n=1).

Chapter 4

IV . Discussion
IV.1 Socio-demographic characteristics

The mean age of our population was 68.3 ±10.5 with a male predominance. Almost half of the patients had severe COPD at NYHA stage 3 and 4 in agreement with another study [16].

IV.2 Methodology

Sputum cytobacteriological examination (SCE) is a widely used non-invasive diagnostic technique. In order to be diagnostically useful, the specimen must meet cytological quality criteria: less than ten epithelial cells and more than 25 polynuclear cells per microscopic field at low magnification. Samples that do not meet these criteria should not be analysed further. The main disadvantage of this technique is the difficulty in obtaining a good quality sample, which is accentuated in the elderly who are often unable to produce sputum of sufficient quality. Interpretation of the ECBC should take into account both the predominant bacterial morphotype(s) on Gram stain and the predominant bacterial species on quantitative culture (pure culture or more than 10^7 CFU per millilitre). When the sample obtained is of good quality, the sensitivity and specificity of the Gram stain vary from 57 to 82% and from 93 to 97% for pneumococcus, respectively. Sensitivity and specificity are 79% and 96% for *Haemophilus spp.* pneumonia, 76% and 96% for staphylococcal pneumonia, 78% and 95% for Gram-negative bacilli pneumonia (other than *Haemophilus spp.*) [17]. When ECBC is possible, it is not always easy to distinguish bacterial colonisation from infection because the oropharynx is frequently colonised by gram-negative bacilli in the elderly. This bacterial flora of the upper airways may vary according to the antibiotic pressure received, any previous hospitalisations and the microbial ecology of the living environment. The ECBC is therefore a low yield test. However, it can be an aid to antibiotic therapy when it reveals the presence of multi-resistant bacteria or bacteria not covered by the probabilistic treatment in place.

A limitation in the interpretation of sputum cultures is that it is not always possible to distinguish between pre-existing colonisation and the acquisition of a new infecting strain. While the bronchial tubes of a healthy individual are usually sterile, the bronchial tubes of a COPD patient are frequently colonised with *H. influenzae, S. pneumoniae* (pneumococcus) or *B. catarrhalis*, even outside of exacerbations. In addition, in COPD, the indications and modalities for sputum cytobacteriological analysis vary from one recommendation to another. According to the SPLF, ECBC is not recommended as a first-line test because of the frequency of false negatives and false positives due to pharyngeal contamination. In case of clinical failure, especially if there are risk factors for *P. aeruginosa* infection, ECBC seems useful.

However, this test has been recommended by the ATS and ESR for any patient hospitalised with EABPCO [18].

For outpatients, this test will only be performed in patients with recent prior antibiotic therapy. These limitations can be overcome by strain differentiation using new molecular approaches such as PCR for pathogen detection. However, due to their complexity, these molecular approaches are not applicable to routine clinical practice. The use of a rapid diagnostic test for atypical viruses and bacteria was able to confirm the presence of these pathogens in a large proportion of cases [19].

IV.3. Prediction of bacterial infection

The Anthonisen criteria (worsening of dyspnoea, increase in sputum quantity and increase in sputum purulence) have long been used. The presence of two or three criteria was considered indirect evidence of a positive culture [20]. A strategy for antibiotic therapy is to select patients, classify them into groups and then propose an antibiotic or list of antibiotics for each group. This selection is based on the Anthonisen classification. The first rule of this selection is that patients with only one criterion of the Anthonisen classification should not receive an antibiotic and only patients with 2 or 3 criteria can benefit from it [21]. Indeed, one study has shown that sputum purulence appears to be the best predictor of bacterial infection [22]. However, differentiating between mucous and purulent sputum alone is not sufficient to identify patients who may benefit from antibiotic therapy [23]. Therefore, cytobacteriological examination of sputum has been recommended by the American Thoracic Society and the European Respiratory Society (ATS/ESR) for all patients hospitalised with CAPCO. However, the rate of positive cultures of spontaneous sputum from our included patients was low

(16.5%) compared to another study that found a higher rate of 32.3% [24]. This can be explained by the presence of several insignificant cultures with polymorphic flora in our samples. However, it is important to take into account the choice of the most protected sampling methods, such as tracheal aspiration in the case of pneumopathy or intubated patients, as well as to respect the positivity thresholds adopted, including a threshold of 10^7 cfu/ml for the quantitative bacteriology of sputum in accordance with French recommendations. However, in our case, intubated patients were excluded from the study, so we were obliged to collect sputum from spontaneous sputum.

The role played by bacteria is also problematic. In the absence of signs and symptoms of infection, the mechanisms underlying the recovery of the low PPM load from bronchial secretions in these patients are open to debate, as it may be related to oropharyngeal bacteria that have migrated to the bronchial tree or to flora colonising the lower bronchial tree.

The low rate of positive cultures found in our study **(Table X)** is acceptable because in most international publications, the culture of a pathogen is considered significant, regardless of the number of colonies identified, as long as the sample meets the cytological criteria of Murray and Washington [25].

Table X. Comparative frequency (%) of potentially pathogenic bacterial species isolated during acute exacerbations of chronic obstructive pulmonary disease in different studies.

Potentially pathogenic bacteria	China 2007-2008 (n=87) [52]	Spain 2009-2010 (n=161) [53]	India 2012 (n=50) [54]	Turkey 2011-2013 (n=242) [55]	Tunisia 2013-2015 (n=240)
% positive culture	54%	16.4%	42%	45%	16.4%
Klebsiella pneumoniae	6.9 % (n=5)	-	33.3% (n=7)	4.8 % (n=2)	16.2 % (n=5)
Pseudomonas aeruginosa	10.3 % (n=6)	30.7% (n=27)	19% (n=4)	21.4 % (n=9)	25.8 % (n=8)
Escherichia coli	3.4 % (n=2)	-	9.5 % (n=2)	9.5 % (n=4)	9.7% (n=3)
Acinetobacter spp.	-	-	9.5 % (n=2)	16.7 % (n=6)	3.2 % (n=1)
Staphylococcus aureus	-	-	14.3 % (n=3)	-	3.2 % (n=1)
Haemophilus influenzae	6.9 % (n=6)	15.9 % (n=14)	-	-	13% (n=4)
Streptococcus pneumoniae	10.3% (n=6)	26.1% (n=23)	-	-	-

In our microbiology laboratory, bacteria are identified according to the European standard recommendations of the French Microbiology Society [15]. A culture was significant when the dominant flora exceeded the quantitative threshold of 10^7 CFU per millilitre. However, we performed these cultures on sputum with a composition of epithelial cells (<10/field) and leukocytes (>25/field) in accordance with the recommendations **(Table I)**. When this is not the case, the sample is repeated.

IV.4. Atypical germs

In general, atypical infections are not identified in the diagnosis because the aetiology of respiratory infections has only been investigated in a small proportion of patients, for those who do not respond to treatment with unresponsiveness to conventional antimicrobial therapy or in cases of severe pneumonia [8].

In contrast to potentially pathogenic microorganisms (PPM), data on the aetiology of atypical germs in EABPCO are controversial [9]. In our study, these atypical organisms are important in the aetiology of respiratory infections in EABPCO, such as *M. pneumoniae* (9%), *C. pneumoniae* (8.4%), and *C. burnetii* (6.6%) which was in agreement with the results of a study that showed an incidence of *C. pneumoniae* infection of 4% to 16% [26]. However, these detections were based exclusively on serology. A recent study [27] using a combination of serology and molecular biology (PCR) in a population of 92 patients with EABPCO showed a prevalence of 2.2% and 4.3% of *M. pneumoniae* and *C. pneumoniae* infections respectively. *L. pneumophila* was conspicuous by its absence in our list of atypical bacteria found. However, a study on the aetiology of EABPCO [6] showed a prevalence of this bacterium in 16.7% of cases.

The European Respiratory Society (ERS) guidelines recognise the atypical pathogens involved in EABPCO, in particular *M. pneumoniae* and *C. pneumoniae*, where macrolides, fluoroquinolones and tetracyclines are more effective. Considering the frequency of coinfections found in our specimens, these aetiologies should be considered in the therapeutic management of patients who fail to respond to initial therapy.

In our study, these were only cases of EABPCO, ideally these patients should have been matched on common characteristics admitted for another non-respiratory condition and stable COPD patients to be able to differentiate. Giving the prevalence of positive serology does not answer the question of the role of these bacteria in COPD. Indeed, *M. pneumoniae* and *C. pneumoniae* infections were often described as a source of failure of previous treatments, in particular betalactam antibiotics, as indicated in our case, favouring their admission to hospital. Hence this selection bias would favour the incidence rate of these atypical bacteria.

IV.5. Potentially pathogenic microorganisms
Table X shows a comparison of the frequency of potentially pathogenic bacterial species isolated during CAPCO in different studies with the results of our study. Taking into account these results, the clinical impact of the management of these patients justifies the tendency of physicians to use other antibiotics such as the 3^{éme} generation fluoroquinolones to avoid this emerging existence.

The importance of bacteriological diagnosis of respiratory infections is based on the fact that the presence of bacteria isolated from the lower respiratory tract such as Gram-negative bacilli, including *Pseudomonas aeruginosa, Escherichia coli* and *Proteus mirabilis* could reflect repeated selective pressure to antibiotic exposure, a greater degree of pulmonary compromise, or a combination of these factors.

In our study, *K. pneumoniae* was isolated as frequently as *H. influenzae* and *B. catarrhalis*. The presence of ESBL-producing strains of *K. pneumoniae* makes it necessary to continuously monitor the susceptibility of the flora of patients hospitalised with CAPD and to regularly re-evaluate first-line antibiotic therapy protocols. In general, guidelines should focus primarily on antibiotic coverage of *H. influenzae, B. catarrhalis* and *S. pneumoniae*, but also against atypical respiratory pathogens.

A limitation in the interpretation of sputum cultures is that it is not always possible to distinguish between pre-existing colonisation and the acquisition of a new infecting strain. While the bronchial tubes of a healthy individual are usually sterile, the bronchial tubes of a COPD patient are frequently colonised with *H. influenzae, S. pneumoniae* (pneumococcus) or *B. catarrhalis*, even outside of exacerbations. In addition, in COPD, the indications and modalities for sputum cytobacteriological analysis vary from one recommendation to another. This test is not routinely recommended and is only indicated in certain situations of failure. In 2004, the ATS/ERS experts suggested that ECBC should be performed in all patients hospitalised for the episode of COPD. For patients managed as outpatients, this test should only be performed in patients with recent prior antibiotic therapy. These limitations can be overcome by strain differentiation using new molecular approaches to pathogen detection such as polymerase chain reaction (PCR). However, due to their complexity, these molecular approaches are not applicable to routine clinical practice.

IV.6. Antibiotic therapy and bacterial resistance

Early initiation of antibiotic treatment is the cornerstone for treating bacterial infection and is associated with improved clinical outcome of patients [28; 29]. On the other hand, exposure to antibiotics leads to selection pressure with a risk of emergence of bacterial resistance and potential adverse effects, sometimes at a significant cost [30]. Indeed, non-fermentative Gram-negative bacilli such as *P. aeruginosa* and *Acinetobacter baumannii* have been reported to be multi-resistant (multi-resistant bacteria or MRB) [11]. These bacteria are most common in patients who have had previous courses of antibiotics, especially when these were inappropriate. Therefore, the use of antibiotics in the treatment of COPD should have three objectives: to hasten the regression of the symptoms of the acute exacerbation, to avoid deterioration of lung function and to delay the occurrence of the next exacerbation as long as possible. Although their efficacy is based on a low level of evidence, they can reduce the risk of treatment failure and death, provided that the local prevalence of resistance and previous antibiotic therapy received by the patient are taken into account. Most current guidelines recommend the prescription of antibiotics for COPD.

In current practice, this antibiotic therapy is routinely prescribed. It is perhaps surprising that the issue of antibiotics in CAPCO has been a hotly debated topic to date. The evidence is gradually accumulating that routine antibiotic therapy is unnecessary. Indeed, the criterion for antibiotic therapy is based on the absence of bacteria identified by current techniques in about 50% of COPD exacerbations, suggesting that antibiotic treatment should not be routine. Results from placebo-controlled antibiotic studies indicate that the benefit of antibiotic treatment is limited to patients with Anthonisen type 1 or 2 exacerbations [31].

ECBC is not recommended as a first-line test because of the frequency of false negatives and false positives from pharyngeal contamination. In case of clinical failure, especially if there are risk factors for *P. aeruginosa* infection, ECBC seems useful [32].

For years a strategy of antibiotic therapy has been designed which consists of selecting patients, classifying them into groups and then proposing an antibiotic or a list of antibiotics for each group. This selection is based on the Anthonisen classification combined with another classification proposed by the Canadian Thoracic Society. The latter distinguishes COPD into three classes according to severity: uncomplicated, complicated and at risk for *Pseudomonas* superinfections [33] **(Table XI)**.

Table XI. Classification of COPD exacerbations [33].

Patient characteristics	uncomplicated	complicated	Risk of *Pseudomonas*
Age	All ages	>65 years	>65 years
Exacerbations (n/year)	<4	>4	>4
Comorbidity	no	yes	Yes
FEV1	>50%	<50%	<50%
Other			Chronic bronchial infections
			Frequent use of antibiotics
			Treatment with corticosteroids

Macrolides or the new cephalosporins were the antibiotics of choice for uncomplicated CAPCO. For complicated CAPD, the antibiotics of choice were amoxicillin/clavulanic acid and the new fluoroquinolones, led by Levofloxacin.

An important point has been pointed out since the recommendations of the AFSSPS, SPILF and SPLF, anti-pneumococcal fluoroquinolones (FQAP) should not be prescribed if the patient has received a fluoroquinolone, whatever the indication, in the last 3 months. It is recommended that they be used with caution in institutions (risk of transmission of resistant strains) and in elderly patients undergoing systemic corticosteroid therapy (increased risk of tendinopathy). Ciprofloxacin, the only anti-pyocyanidant that can be administered orally, was the antibiotic of choice in the case of a risk of pyocyanic infection [34].

Treatment with Amoxicillin-Clavulanic acid or second generation macrolides or the new quinolones seems to be most appropriate and effective in the group of patients with the most severe COPD whose baseline obstructive syndrome is severe requiring mechanical ventilation where bacterial infection seems to play an important role.

In these patients, the use of conventional antibiotics such as Amoxicillin is not recommended due to the high level of bacterial resistance [35].

In a prospective randomised controlled double-blind study [36], Nouira et al. showed the benefit of the new fluoroquinolones in the treatment of CAPCO. Levofloxacin. The latter currently has the most stable activity against pneumococcus and *Haemophilus* with a resistance rate approaching 0% in the majority of countries where it has been evaluated, as is the case with the results of our study. However, it would be obvious that there is still insufficient hindsight to have a clear picture on this issue. A significant rate of resistance to the combination Amoxicillin-Clavulanic acid was observed in the strains isolated in our work. While this combination is still effective against pneumococcus and *Haemophilus*.

IV.6.1. Enterobacteriaceae and antibiotic resistance

Enterobacteriaceae are the most common class of BGN that cause diseases of widely varying severity, due to distinct pathogenic mechanisms, they have acquired the ability to produce diverse resistance mechanisms that present a great threat to the future through its impact on morbidity and mortality. They are capable of producing beta-lactamases, enzymes that inactivate the major class of antibiotics, the beta-lactams, by opening the beta-lactam cycle. Although the first line of antibiotic therapy is amoxicillin or amoxicillin/clavulanic acid, the major finding in our study was the resistance of all our isolated strains (*E. coli, K. pneumoniae, P. mirabilis* and *Providencia spp*) to betalactam. In addition, all *K. pneumoniae* species isolated were extended-spectrum beta-lactamase (ESBL) producers except for one low-level penicillinase producer. Over the past 30 years, *K. pneumoniae has* developed enzymes resistant to these beta-lactams. The first ESBL enzyme capable of blocking cephalosporins was described in 1985. Then, a beta-lactamase attacking imipenem was isolated in Greece in 2003 from this same bacterial species. In our study, all isolated Enterobacteriaceae were sensitive to carbapenem (imipenem) which was comparable to most European countries where the proportion of carbapenem resistant strains was less than 1%. In 2010, an increasing trend of carbapenem-resistant *K. pneumoniae was* observed for Austria, Cyprus, Hungary and Italy. This increasing trend is a particularly worrying phenomenon as carbapenems are the last active antibiotics for the treatment of multi-drug resistant Gram-negative infections including those producing ESBL *(Ministry of Health, 2010).* A favourable evolution under antibiotic therapy in relation to the resistance profile of isolated germs (*P. aeruginosa* and *K. pneumoniae* ESBL) can be explained by carriage, it is a colonisation rather than an infection

IV.6.2. Non-enterobacteria and antibiotic resistance

The pyocyanine bacillus is known to have a high level of natural resistance to most antibiotics against gram-negative bacteria, such as the aminopenicillins, cephalosporins 1re , 2ème and 3ème generation (cefotaxime, ceftriaxone), the older fluoroquinolones, but also tetracyclines and cotrimoxazole. The natural resistance of *P. aeruginosa is* characterised by its ability to develop acquired resistance to almost all antibiotic molecules available in therapeutics and to which it is exposed. In our study, we found that *P. aeruginosa* has significant resistance to carboxypenicillins (Ticarcillin with or without clavulanic acid), fosfomycin and 3ème generation cephalosporins. These figures help to explain the high number of inadequate empirical antibiotic treatments for *P. aeruginosa* infections. In France, a review of the literature confirmed that imipenem, ceftazidime, piperacillin-tazobactam, tobramycin and amikacin are the most regularly active molecules in vitro on this bacterium with activity rates ranging from 64% to 87% [37]. These data are in line with our results in which we found that *P. aeruginosa* shows sensitivity to imipenem, amikacin, piperacillin-tazobactam, gentamicin with a resistance rate ranging from 14.2 to 37.5%, and only colistin shows a resistance percentage of 0%. This makes colistin the only antibiotic active against *P. aeruginosa.*

A. baumannii is a bacterium that is frequently resistant to many antibiotics and is responsible for epidemics of pulmonary infections. It can persist for a long time in the hospital environment and its transmission is manuported. The *A. baumannii* strain was first identified in the North of France in July 2001. It has acquired antibiotic resistance characteristics that make it a cause for concern but facilitate its identification. It produces an enzyme (extended spectrum beta-lactamase, or ESBL, type VEB-1) which makes it resistant to all beta-lactams; indeed, the rate of resistance of *A. baumanii* found in our study is 100% to all beta-lactams. Only colistin and amikacin have a better activity against *A. baumanii* with a sensitivity of 100% and 50% respectively. In fact, colistin remains the only antibiotic active against this multi-resistant germ. The percentage of resistance to imipenem has reached 100%, which poses a real therapeutic problem. In a retrospective study carried out in 2010 at the Tahar Sfar Hospital in Mahdia, the rate of resistance to imipenem was 31% [38]. Among the aminoglycosides, amikacin remains the only most active molecule, as noted in our study; *A. baumannii* presents a fairly high rate of resistance to gentamicin (100%) compared to amikacin (50%).

According to our study, fluoroquinolones have lost much of their effectiveness against *A. baumannii* (100% for ciprofloxacin). In developed countries, this rate did not exceed 25%. However, the new fluoroquinolones such as levofloxacin would be more active against *A. baumannii* than ciprofloxacin [39].

IV.6.3. Staphylococci and antibiotic resistance

Anti-staphylococcal antibiotic therapy aims to rapidly decrease the bacterial inoculum without selecting resistant mutants. Due to the pharmacodynamic properties of currently available antibiotics and the resistance rates, antibiotic combinations were often prescribed for severe infections. In our study, 50% of strains were resistant to penicillin G due to their production of a penicillinase that hydrolyses penicillin G, amino-, carboxy- and ureidopenicillins [40]. This level of resistance is comparable to that found in the study by *Batard et al*, where approximately 95% of strains were resistant to penicillin G [41].

IV.6.4. Pneumococcus and antibiotic resistance

The emergence of multi-drug resistant strains of pneumococcus has been a concern for several years. Since the isolation of the first strain in Australia in 1967, pneumococcal resistance has been increasing [42]. In a Tunisian study, the proportions of strains with decreased susceptibility to penicillin G (PSDP), amoxicillin and cefotaxime were 48%, 28% and 19% respectively [43]. And in a study conducted in 1996 and 1997, which compared data from nine African countries, Tunisia was among the countries with a high level of penicillin resistance (the second country after Senegal). Given the small number of strains isolated in our study (N=3), these proportions were all at 0%. However, the rate of resistance remains low and these antibiotics still have a place in therapy [44].

In our study, this was a severe COPD population according to the GOLD classification (with an average of 3 exacerbations in the year preceding the study). Some of them probably received recent antibiotic therapy. This epidemiological study on antibiotic resistance also suggests that the prescription of first-line antibiotics for these patients should be reviewed, particularly for the combination amoxicillin-clavulanic acid.

Finally, one of the main orientations to come must be focused on the possibility of targeting patients to be treated, less by standard bacteriological examinations (aspiration, sputum examination) which have demonstrated their limits than by new markers of infection and inflammation (CRP for example) and other new techniques.

IV.7. Prediction of bacterial infection

An antibiotic therapy strategy consists of selecting patients, classifying them into groups and then proposing an antibiotic or a list of antibiotics for each group. This selection has long been based on the Anthonisen classification. The Anthonisen criteria (worsening of dyspnoea, increase in sputum quantity and increase in sputum purulence) have long been used for the classification of EABPCO [45]. Taking into account the severity of COPD as a comorbidity and the Anthonisen classification seems reasonable [46]. The first rule pointed out that patients with only one criterion of Anthonisen's classification should not receive antibiotics and that only patients with two or three criteria can benefit [47]. The presence of two or three criteria was considered indirect evidence of a positive culture [48]. This classification depends on the number of cardinal symptoms and the presence of certain accompanying symptoms as shown in **Table XII**.

Table XII. Anthonisen classification of EABPCO based on cardinal symptoms. According to data from Anthonisen and colleagues (1987)

Severity of exacerbation	Type of exacerbation	Characteristics
Severe	Type 1	Increased dyspnea, sputum volume and sputum purulence
Moderate	Type 2	Any 2 of the above 3 cardinal symptoms
Mild 3	Type 3	Any 1 of the above 3 cardinal symptoms and 1 or more of the following minor symptoms or signs

		- Cough
		- Wheezing
		- Fever without an obvious source
		- Upper respiratory tract infection in the past 5 days
		- Respiratory rate increase 20% over baseline
		- Heart rate increase 20% over baseline

Despite its simplicity, this Anthonisen classification is not without subjectivity. It does not apply to all patients, especially those requiring mechanical ventilation. These patients should therefore at least be considered as type 1 patients (even in the absence of the two sputum criteria) [46].

Other criteria predictive of bacterial infection have been less well studied. According to Afssps, fever lasting more than four days would make the probability of a viral infection low and point to a bacterial cause. In addition, a study has shown that sputum purulence seems to be the best predictor of bacterial infection [49]. However, differentiating between mucous and purulent sputum is not sufficient to identify patients who may benefit from antibiotic therapy [50]. Finally, all proposed strategies should be validated and evaluated.

Chapter 5

V. Conclusion

The present study devoted to the bacterial flora of EABPCO, showed a low positivity of the quantitative bacteriology of sputum, a significant percentage of resistant strains 40.9% with a predominance of pseudomonas which are exclusively multi-resistant. These results may help in the management of these patients allowing physicians to better manage the use of antibiotics, prescribe adequate drugs, and institute adequate isolation of infected patients in order to minimize the effect of infection and nosocomial transmission.

Prospectively, in order to effectively control recurrent exacerbations and readmission to hospitals of patients with re-exacerbations of COPD, and thus reduce the use of antibiotic therapy, studies can be done that take into account the genetic diversity between individual strains of a bacterial species, including changes in surface antigenic structures. Such variation in the surface antigenic structure of bacterial pathogens allows these organisms to evade pre-existing host immunity and cause recurrent infection.

Acknowledgements

Many thanks to the emergency physicians of the participating centres who participated in the good recruitment of patients and to the technicians of the microbiology laboratory of the Fattouma Bourguiba University Hospital of Monastir for their help.

Many thanks also to all the patients who agreed to participate in the study.

Study budget

This work was supported by the Fattouma Bourguiba University Hospital (Monastir, Tunisia) for the performance of biological analyses. No additional budget was required.

Conflict of interest

No conflict of interest.

References

1. Ball P: Epidemiology and treatment of chronic bronchitis and its exacerbations. *CHEST Journal* 1995; 108(2_Supplement): 43S-52S.

2. Lopez-Campos JL, Ruiz-Ramos M and Soriano JB: Mortality trends in chronic obstructive pulmonary disease in Europe, 1994-2010: ajoin point regression analysis. *The Lancet RespiratoryMedicine* 2014; 2(1): 54-62.

3. Anthonisen NR et al: Antibiotic therapy in exacerbations of chronic obstructive pulmonary disease. *Annals of internal medicine* 1987; 106(2): 196-204.

4. Fanny W S Ko, Margaret Ip, Paul K S Chan, Michael C H Chan, Kin-Wang To, Susanna S S Ng, Shirley S. L. Chau, Julian W. Tang, and David S. C. Hui: Viral etiology of acute exacerbations of COPD in Hong Kong. *CHEST* 2007; 132:900-8.

5. Birolleau S: GENERAL REVIEW Treatment of exacerbation of chronic obstructive pulmonary disease (COPD). *J Fran Viet Pneu* 2012; 03(07): 1-45.

6. Tuuminen T, Varjo S, Ingman H, Weber T, Oksi J and Viljanen M: Prevalence of *Chlamydia pneumoniae* and *Mycoplasma pneumoniae* immunoglobulin G and A antibodies in a healthy Finnish population as analyzed by quantitative enzyme immunoassays. *Clin. Diagn. Lab. Immunol* 2000; 7:734-738.

7. Rosell A, Monso E, Soler N, Torres F, Angrill J, Riise G, Zalacam R, Morera J, and Torres A: Microbiology determinants of exacerbation in chronic obstructive pulmonary disease. *Arch Intern Med* 2005; 165:891-897.

8. Kumar S and Hammerschlag MR: Acute respiratory infection due to *Chlamydia pneumoniae*: current status of diagnostic methods. *Clin Infect Dis* 2007; 44:568-576.

9. Lieberman D, Lieberman D, Ben-Yaakov M, Shmarkov O, Gelfer Y, Varshavsky R, Ohana B, Lazarovich Z and Boldur I: Serological evidence of *Mycoplasma pneumoniae* infection in acute exacerbation of COPD. *Diagn Microbiol Infect Dis* 2002; 44: 1-6.

10. Abdallah FCB, Taktak S, Chtourou A, Mahouachi R and Kheder AB: burden of chronic respiratory diseases (CRD) in Middle East and North Africa (MENA). *World Allergy Organ J2011*;4(1),1.

11. Nseir S and Ader F: Prevalence and outcome of severe chronic obstructive pulmonary disease exacerbations caused by multidrug-resistant bacteria. *Curr Opin Pulm Med* 2008; 14(2): 95-100.

12. Fisher J D: New York Heart Association Classification. *Arch Intern Med* 1972; 129(5), 836.

13. Nouira S, Marghli S, Belghith M, Besbes L, Elatrous S and Abroug F: Once daily oral ofloxacin in chronic obstructive pulmonary disease exacerbation requiring mechanical ventilation: a randomised placebo-control trial. *Lancet* 2001; 358: 2020-5.

14. Global initiative for chronic obstructive lung disease (GOLD). Global strategy for the diagnosis, management, and prevention of chronic obstructive pulmonary disease (updated 2016), http://goldcopd.org/global-strategy- diagnosis-management-prevention-copd-2016/ (accessed on the 11th of July 2016).

15. French Society of Microbiology. Bronchopulmonary infections (excluding tuberculosis and 197 cystic fibrosis). In REMIC, Société Française de Microbiologie Ed 2015, pp184-19

16. Dizdari E, Bridevaux P.-O and Leuenberger P. Role of bacteria in COPD exacerbations. *Rev Med Suisse* 2004;24137.

17. Anevlavis S, Petroglou N, Tzavaras A et al. Prospective study of the diagnostic utility of sputum Gram stain in pneumonia. *J Infect*. 2009; 59(2):83-89.

18. Celli BR, MacNee WATS, Agusti AATS et al. Standards for the diagnosis and treatment of patients with COPD: a summary of the ATS/ERS position paper. *Eur Respir J*. 2004;23(6):932-946.

19. Pierce VM, Elkan M, Leet M, McGowan KL, Hodinka RL. Comparison of the Idaho Technology FilmArray system to real-time PCR for detection of respiratory pathogens in children. *J Clin Microbiol*. 2011;JCM- 05996.

20. Murphy TF and Sethi S: Chronic obstructive pulmonary disease. Role of bacteria and guide to antibacterial selection in the older patient. *Drugs Aging* 2002; 19: 761-75.

21. Buisson CB: Antibiotic therapy in COPD exacerbations: a treatment for accepting uncertainty? *Rev Mal Respir* 2004; 21(2): 241-244.

22. Burgel PR. Indications and choice of antibiotic therapy for an exacerbation of chronic obstructive pulmonary disease (COPD). *Med Mal Infect* 2006; 36: 706-17.

23. Brusse-Keizer MG, Grotenhuis AJ, Kerstjens HA, Telgen MC, van der Palen J and Hendrix MG: Relation of sputum color to bacterial load in acute exacerbations of COPD. *Respir Med* 2009; 103: 601-6.

24. Fanny W.S. Ko, Margaret Ip, Paul K.S. Chan, Joan P.C. Fok, Michael C.H. Chan, Jenny C. Ngai, Doris P.S. Chan and David S.C. Hui: A 1-year prospective study of the infectious etiology in patients hospitalized with acute exacerbations of COPD. *CHEST* 2007; 131: 44-52.

25. Leroy O: Contribution of microbiological investigations to the diagnosis of lower respiratory tract infections. *Med Mal Infect* 2006; 36: 570-98.

26. Blasi F, Legnani D, Lombardo V.M, Negretto G.G, Magliano E, Pozzoli R , Chiodo F, Fasoli A and Allegra L : *Chlamydia pneumoniae* infection in acute exacerbations of COPD. *Eur Respir J* 1993: 6: 19-22.

27. Nakou A, Papaparaskevas J, Diamantea F, Skarmoutsou N, Polychronopoulos V and Tsakris A: A prospective study on bacterial and atypical etiology of acute exacerbation in chronic obstructive pulmonary disease. *Future microbiology* 2014; 9(11): 1251-1260.

28. Dellinger RP, Levy MM, Rhodes A et al. Surviving Sepsis Campaign: international guidelines for management of severe sepsis and septic shock, 2012. *Intensive Care Med.* 2013;39(2):165-228.

29. Kumar A, Ellis P, Arabi Y, et al. Initiation of inappropriate antimicrobial therapy results in a fivefold reduction of survival in human septic shock. *CHEST* .2009;136(5):1237-1248.

30. Dennesen PJ, van der VEN AJ, Kessels AG, Ramsay G, Bonten MJ. Resolution of infectious parameters after antimicrobial therapy in patients with ventilator-associated pneumonia. *Am J Respir Crit Care Med.*2001;163(6):1371-1375.

31. Burgel P. R. Indications and choice of antibiotic therapy for an exacerbation of chronic obstructive pulmonary disease (COPD). *Med Mal Infect* 2006; 36(11), 706-717.

32. Society of pulmonology of French language. Recommendations for the management of COPD. Exacerbations-decompensations: antibiotic therapy. *Rev Mal Respir* 2003; 20:4S65-4S68.

33. Balter M. S, Hyland R. H, Low D. E, Renzi P. M, Braude, A. C, and Cole, P. J. Recommendations on the management of chronic bronchitis. *Can Med Assoc J* 1994;151 (Suppl 10):5-23.

34. Parameswaran G. I and Sethi S. Pseudomonas infection in chronic obstructive pulmonary disease. *Future Microbiol* 2012, 7(10), 1129-1132.

35. Nouira S, Marghli S, and Abroug F. Antibiotic therapy and exacerbation of chronic obstructive pulmonary disease. *Réanim* 2003, 12(1), 46-52.

36. Nouira S, Marghli S, Makhlouf B, Besbes L, Elatrous S and Abroug F. Once daily ofloxacin in chronic obstructive pulmonary disease exacerbation requiring mechanical ventilation: a randomised placebo

controlled trial. *Lancet* 2001; 358:2020.

37. Cavallo JD, Hocquet D, Plesiat P, Fabre R, Roussel-Delvallez M. Susceptibility of Pseudomonas aeruginosa to antimicrobials: a 2004 French multicentre hospital study. *JAntimicrob Chemother.* 2007;59:1021-1024.

38. Ben Haj Khalifa A, Khedher M. Antibiotic susceptibility profile of Acinetobacter baumannii strains isolated in the Mahdia region. *Med Mal Infect.* 2010;40:126-128.

39. Heinemann B, Wisplinghoff H, Edmond M, Seifert H. Comparative activities of ciprofloxacin, clinafloxacin, gatifloxacin, gemifloxacin, levofloxacin, moxifloxacin, and trovafloxacin against epidemiologically defined Acinetobacter baumannii strains. *Antimicrob Agents Chemother.* 2000;44: 2211-3.

40. Yves LL, Michel G. Staphylococcus aureus. Editions TEC & DOC. Paris: Lavoisier, 2009.

41. Batard É, Ferron-Perrot C, Caillon J, Potel G. Antibiotic therapy of infections caused by Staphylococcus aureus. *Therapeutic Medicine.* 2005;11(6):395-403.

42. Decastro N, Molina J. Lower respiratory infections in adults. EMC (Elsevier Mason SAS, Paris), pneumology. 2011;6-003-D-10.

43. Znazen A, Ayadi S, Mnif B, et al. Antibiotic resistance of Streptococcus pneumoniae in Tunisia: a multicenter study 2004-2006. *Rev Tun Infect.* 2010;4:10-14.

44. Vergnaud M, Bourdon S, Brun M, et al. Regional pneumococcal observatories: analysis of antibiotic resistance and serotypes of *Streptococcus pneumoniae* in 2001. *BEH.* 2003;37:173-176.

45. Siddiqi A, Sethi S. Optimizing antibiotic selection in treating COPD exacerbations. *Int J Chron Obstruct Pulmon Dis.* 2008;3(1):31.

46. Nouira S, Marghli S, Abroug F. Antibiotic therapy and exacerbation of chronic obstructive pulmonary disease. *Resuscitation.* 2003;12(1):46-52.

47. Buisson CB. Antibiotic therapy in COPD exacerbations: a treatment for accepting uncertainty? *Rev Mal Respir.* 2004;21(2):241-244.

48. Murphy TF, Sethi S. Chronic obstructive pulmonary disease. *Drugs Aging.* 2002;19(10):761-775.

49. Burgel PR. Indications and choice of antibiotic therapy for an exacerbation of chronic obstructive pulmonary disease (COPD). *Med Mal Infect.* 2006;36(11):706-717.

50. (Brusse-Keizer et al., 2009). Brusse-Keizer MGJ, Grotenhuis AJ, Kerstjens HAM et al. Relation of sputum

color to bacterial load in acute exacerbations of COPD. *Respir Med.* 2009;103(4):601-606.

51. Yang X, Strobel M, Tian L, Barennes H, and Buisson Y. Bacterial flora of acute exacerbations of chronic obstructive pulmonary disease (COPD) in Kunming, China. *Med Mal Infect* 2011; 41(4), 186191.

52. Boixeda R, Almagro P, Diez-Manglano J, Cabrera F. J, Recio J, Martin-Gamdo I, and Soriano J. B. (2015). Bacterial flora in the sputum and comorbidity in patients with acute exacerbations of COPD. *Int J Chron Obstruct Pulmon Dis* 2015; 10, 2581.

53. Basu S, Mukherjee S and Samanta A. Epidemiological study of bacterial microbiology in AECOPD patients of Kolkata, India. *Asian J Pharm Clin Res* 2013; 6(1), 112-116.

54. Dilektasli A. G, Cetinoglu E. D, Ozturk N. A. A, Coskun F, Ozkaya G, Ursavas A, Ozakin C, Karadag M, and Uzaslan, E. Bacterial etiology in acute hospitalized chronic obstructive pulmonary disease exacerbations. *EuRJ* 2016;2(2):99-106.

yes I want morebooks!

Buy your books fast and straightforward online - at one of world's fastest growing online book stores! Environmentally sound due to Print-on-Demand technologies.

Buy your books online at
www.morebooks.shop

Kaufen Sie Ihre Bücher schnell und unkompliziert online – auf einer der am schnellsten wachsenden Buchhandelsplattformen weltweit! Dank Print-On-Demand umwelt- und ressourcenschonend produziert.

Bücher schneller online kaufen
www.morebooks.shop

info@omniscriptum.com
www.omniscriptum.com

Printed by Books on Demand GmbH, Norderstedt / Germany